HCF

I0479130

THINGS YOU SHOULD KNOW
(QUESTIONS AND ANSWERS)

By Rumi Michael Leigh

Introduction

I would like to thank you for purchasing this book, *"HCF", things you should know (questions and answers)"*.

This book will help you understand, revise, and have a good general knowledge and understanding of the basics of HCF.

I hope you enjoy it!

Table of Contents

Part 1

1.

What is HCF?

HCF is the Highest Common Factor

Part 2a: HCF of two numbers (Questions)

Find the HCF of the following numbers

a. 10 and 14

b. 50 and 15

c. 18 and 12

d. 72 and 99

e. 80 and 78

f. 60 and 42

g. 74 and 76

h. 20 and 30

i. 70 and 40

j. 54 and 24

Part 2a: HCF of two numbers (Solutions)

Find the HCF of the following numbers

a.10 and 14

The factors of 10 are 1,2,5,10
The factors of 14 are 1,2,7,14
The highest factor found in both 10 and 14 is 2
So, the HCF is 2

b. 50 and 15

The factors of 50 are 1,2,5,10,25,50
The factors of 15 are 1,3,5,15
So, the HCF is 5

c. 18 and 12

The factors of 18 are 1,2,3,6,9,18
The factors of 12 are 1,2,3,4,6,12
So, the HCF is 6

d. 72 and 99

The factors of 72 are 1,2,3,4,6,8,9,12,18,36,72
The factors of 99 are 1,3,9,11,33,99
So, the HCF is 9

e. 80 and 78

The factors of 80 are 1,2,4,5,8,10,16,20,40,80
The factors of 78 are 1,2,3,6,13,26,78
So, the HCF is 2

f. 60 and 42

The factors of 60 are 1,2,3,4,5,6,10,12,15,20,30,60
The factors of 42 are 1,2,3,6,7,14,21,42
So, the HCF is 6

g. 74 and 76

The factors of 74 are 1,2,37,74
The factors of 76 are 1,2,4,19,38,76
So, the HCF is 2

h. 20 and 30

The factors of 20 are 1,2,4,5,10,20
The factors of 30 are 1,2,3,5,6,10,15,30
So, the HCF is 10

i. 70 and 40

The factors of 70 are 1,2,5,7,10,14,35,70
The factors of 40 are 1,2,4,5,8,10,20,40
So, the HCF is 10

j. 54 and 24

The factors of 54 are 1,2,3,6,9,18,27,54
The factors of 24 are 1,2,3,4,6,8,12,24

So, the HCF is 6

Part 2b: HCF of two

numbers (Questions)

Find the HCF of the following numbers

a. 24 and 16

b. 40 and 18

c. 60 and 70

d. 12 and 24

e. 74 and 30

f. 50 and 20

g. 27 and 96

h. 20 and 15

i. 30 and 68

j. 70 and 16

Part 2b: HCF of two numbers (Solutions)

Find the HCF of the following numbers

a. 24 and 16

The factors of 24 are 1,2,3,4,6,8,12,24
The factors of 16 are 1,2,4,8,16
So, the HCF is 8

b. 40 and 18

The factors of 40 are 1,2,4,5,8,10,20,40
The factors of 18 are 1,2,3,6,9,18
So, the HCF is 2

c. 60 and 70

The factors of 60 are 1,2,3,4,5,6,10,12,15,20,30,60

The factors of 70 are 1,2,5,7,10,14,35,70

So, the HCF is 10

d. 12 and 24

The factors of 12 are 1,2,3,4,6,12

The factors of 24 are 1,2,3,4,6,8,12,24

So, the HCF is 12

e. 74 and 30

The factors of 74 are 1,2,37,74

The factors of 30 are 1,2,3,5,6,10,15,30

So, the HCF is 2

f. 50 and 20

The factors of 50 are 1,2,5,10,25,50

The factors of 20 are 1,2,4,5,10,20

So, the HCF is 10

g. 27 and 96

The factors of 27 are 1,3,9,27
The factors of 96 are 1,2,3,4,6,12,16,24,32,48,96
So, the HCF is 3

h. 20 and 15

The factors of 20 are 1,2,4,5,10,20
The factors of 15 are 1,3,5,15
So, the HCF is 5

i. 30 and 68

The factors of 30 are 1,2,3,5,6,10,15,30
The factors of 68 are 1,2,17,34,68
So, the HCF is 2

j. 70 and 16

The factors of 70 are 1,2,5,7,10,14,35,70
The factors of 16 are 1,2,4,8,16
So, the HCF is 2

Part 2c: HCF of two numbers (Questions)

Find the HCF of the following numbers

a. 74 and 18

b. 60 and 74

c. 72 and 94

d. 50 and 98

e. 30 and 36

f. 70 and 32

g. 40 and 72

h. 90 and 14

i. 20 and 28

j. 72 and 26

Part 2c: HCF of two numbers (Solutions)

Find the HCF of the following numbers

a. 74 and 18

The factors of 74 are 1,2,37,74
The factors of 18 are 1,2,3,6,9,18
So, the HCF is 2

b. 60 and 74

The factors of 60 are 1,2,3,4,5,6,10,12,15,20,30,60
The factors of 74 are 1,2,37,74
So, the HCF is 2

c. 72 and 94

The factors of 72 are 1,2,3,4,6,8,9,12,18,36,72
The factors of 94 are 1,2,47,94
So, the HCF is 2

d. 50 and 98

The factors of 50 are 1,2,5,10,25,50
The factors of 98 are 1,2,7,14,49,98
So, the HCF is 2

e. 30 and 36

The factors of 30 are 1,2,3,5,6,10,15,30
The factors of 36 are 1,2,3,4,6,9,12,18,36
So, the HCF is 6

f. 70 and 32

The factors of 70 are 1,2,5,7,10,14,35,70
The factors of 32 are 1,2,4,8,16,32
So, the HCF is 2

g. 40 and 72

The factors of 40 are 1,2,4,5,8,10,20,40
The factors of 72 are 1,2,3,4,6,8,9,12,18,36,72
So, the HCF is 8

h. 90 and 14

The factors of 90 are 1,2,3,5,6,9,10,15,30,45,90
The factors of 14 are 1,2,7,14
So, the HCF is 2

i. 20 and 28

The factor of 20 are 1,2,4,5,10,20
The factors of 28 are 1,2,4,7,14,28
So, the HCF is 4

j. 72 and 26

The factors of 72 are 1,2,3,4,6,8,9,12,24,36,72
The factors of 26 are 1,2,13,26
So, the HCF is 2

Part 2d: HCF of two numbers (Questions)

Find the HCF of the following numbers

a. 80 and 32

b. 40 and 27

c. 72 and 94

d. 50 and 40

e. 96 and 16

f. 30 and 18

g. 70 and 100

h. 28 and 40

i. 60 and 64

j. 74 and 27

Part 2d: HCF of two numbers (Solutions)

Find the HCF of the following numbers

a. 80 and 32

The factors of 80 are 1,2,4,5,8,10,16,20,40,80
The factors of 32 are 1,2,4,8,16,32
So, the HCF is 16

b. 40 and 27

The factors of 40 are 1,2,4,5,8,10,20,40
The factors of 27 are 1,3,9,27
So, the HCF is 1

c. 72 and 94

The factors of 72 are 1,2,3,4,6,8,9,12,24,36,72
The factors of 94 are 1,2,47,94
So, the HCF is 2

d. 50 and 40

The factors of 50 are 1,2,5,10,25,50
The factors of 40 are 1,2,4,5,8,10,20,40
So, the HCF is 10

e. 96 and 16

The factors of 96 are 1,2,3,4,6,12,16,24,32,48,96
The factors of 16 are 1,2,4,8,16
So, the HCF is 16

f. 30 and 18

The factors of 30 are 1,2,3,5,6,10,15,30
The factors of 18 are 1,2,3,6,9,18
So, the HCF is 6

g. 70 and 100

The factors of 70 are 1,2,5,7,10,14,35,70
The factors of 100 are 1,2,4,5,10,20,25,50,100
So, the HCF is 10

h. 28 and 40

The factors of 28 are 1,2,4,7,14,28
The factors of 40 are 1,2,4,5,8,10,20,40
So, the HCF is 4

i. 60 and 64

The factors of 60 are 1,2,3,4,5,6,10,12,15,20,30,60
The factors of 64 are 1,2,4,8,16,32,64
So, the HCF is 4

j. 74 and 27

The factors of 74 are 1,2,37,74
The factors of 27 are 1,3,9,27

So, the HCF is 1

Part 2e: HCF of two numbers (Questions)

Find the HCF of the following numbers

a. 72 and 16

b. 60 and 12

c. 40 and 55

d. 80 and 16

e. 30 and 22

f. 70 and 15

g. 50 and 64

h. 20 and 16

i. 80 and 14

j. 74 and 12

Part 2e: HCF of two numbers (Solutions)

Find the HCF of the following numbers

a. 72 and 16

The factors of 72 are 1,2,3,4,6,8,9,12,18,36,72
The factors of 16 are 1,2,4,8,16
So, the HCF is 8

b. 60 and 12

The factors of 60 are 1,2,3,4,5,6,10,12,15,20,30,60
The factors of 12 are 1,2,3,4,6,12
So, the HCF is 12

c. 40 and 55

The factors of 40 are 1,2,4,5,8,10,20,40
The factors of 55 are 1,5,11,55
So, the HCF is 1

d. 80 and 16

The factors of 80 are 1,2,4,5,8,10,16,20,40,80
The factors of 16 are 1,2,4,8,16
So, the HCF is 16

e. 30 and 22

The factors of 30 are 1,2,3,5,6,10,15,30
The factors of 22 are 1,2,11,22
So, the HCF is 2

f. 70 and 15

The factors of 70 are 1,2,5,7,10,14,35,70
The factors of 15 are 1,3,5,15
So, the HCF is 5

g. 50 and 64

The factors of 50 are 1,2,5,10,25,50
The factors of 64 are 1,2,4,8,16,32,64
So, the HCF is 2

h. 20 and 16

The factors of 20 are 1,2,4,5,10,20
The factors of 16 are 1,2,4,8,16
So, the HCF is 4

i. 80 and 14

The factors of 80 are 1,2,4,5,8,10,16,20,40,80
The factors of 14 are 1,2,7,14
So, the HCF is 2

j. 74 and 12

The factors of 74 are 1,2,37,74
The factors of 12 are 1,2,3,6,12

So, the HCF is 2

Part 2f: HCF of two

numbers (Questions)

Find the HCF of the following numbers

a. 72 and 58
b. 60 and 15
c. 22 and 16
d. 90 and 10
e. 30 and 12
f. 70 and 15
g. 50 and 56
h. 40 and 100
i. 80 and 15
j. 75 and 100

Part 2f: HCF of two

numbers (Solutions)

Find the HCF of the following numbers

a. 72 and 58

The factors of 72 are 1,2,3,4,6,8,9,12,24,36,72
The factors of 58 are 1,2,29,58
So, the HCF is 2

b. 60 and 15

The factors of 60 are 1,2,3,4,5,6,10,12,15,20,30,60
The factors of 15 are 1,3,5,15
So, the HCF is 15

c. 22 and 16

The factors of 22 are 1,2,11
The factors of 16 are 1,2,8,16
So, the HCF is 2

d. 90 and 10

The factors of 90 are 1,2,3,5,6,9,10,15,30,45,90
The factors of 10 are 1,2,5,10
So, the HCF is 10

e. 30 and 12

The factors of 30 are 1,2,3,5,6,10,15,30
The factors of 12 are 1,2,3,4,6,12
So, the HCF is 6

f. 70 and 15

The factors of 70 are 1,2,5,7,10,14,35,70
The factors of 15 are 1,3,5,15
So, the HCF is 5

g. 50 and 56

The factors of 50 are 1,2,5,10,25,50
The factors of 56 are 1,2,4,7,8,14,28,56
So, the HCF is 2

h. 40 and 100

The factors of 40 are 1,2,4,5,8,10,20,40
The factors of 100 are 1,2,4,5,10,20,25,50,100
So, the HCF is 20

i. 80 and 15

The factors of 80 are 1,2,4,5,8,10,16,20,40,80
The factors of 15 are 1,3,5,15
So, the HCF is 5

j. 75 and 100

The factors of 75 are 1,3,5,15,25,75
The factors of 100 are 1,2,4,5,10,20,25,50,100

So, the HCF is 25

Part 2g: HCF of two

numbers (Questions)

Find the HCF of the following numbers

a. 90 and 15

b. 30 and 14

c. 60 and 75

d. 30 and 24

e. 70 and 12

f. 40 and 50

g. 80 and 62

h. 74 and 28

i. 50 and 32

j. 72 and 16

Part 2g: HCF of two numbers (Solutions)

Find the HCF of the following numbers

a. 90 and 15

The factors of 90 are 1,2,3,5,6,9,10,15,30,45,90
The factors of 15 are 1,3,5,15
So, the HCF is 15

b. 30 and 14

The factors of 30 are 1,2,3,5,6,10,15,30
The factors of 14 are 1,2,7,14
So, the HCF is 2

c. 60 and 75

The factors of 60 are 1,2,3,4,5,6,10,12,15,20,30,60
The factors of 75 are 1,3,5,15,25,75
So, the HCF is 15

d. 30 and 24

The factors of 30 are 1,2,3,5,6,10,15,30
The factors of 24 are 1,2,3,4,6,8,12,24
So, the HCF is 6

e. 70 and 12

The factors of 70 are 1,2,5,7,10,14,35,70
The factors of 12 are 1,2,3,4,6,12
So, the HCF is 2

f. 40 and 50

The factors of 40 are 1,2,4,5,8,10,20,40
The factors of 50 are 1,2,5,10,25,50
So, the HCF is 10

g. 80 and 62

The factors of 80 are 1,2,4,5,8,10,16,20,40,80
The factors of 62 are 1,2,22,31,62
So, the HCF is 2

h. 74 and 28

The factors of 74 are 1,2,37,74
The factors of 28 are 1,2,4,7,14,28
So, the HCF is 2

i. 50 and 32

The factors of 50 are 1,2,5,10,25,50
The factors of 32 are 1,2,4,8,16,32
So, the HCF is 2

j. 72 and 16

The factors of 72 are 1,2,3,4,6,8,9,12,24,36,72
The factors of 16 are 1,2,8,16

So, the HCF is 8

Part 2h: HCF of two

numbers (Questions)

Find the HCF of the following numbers

a. 100 and 45

b. 40 and 54

c. 74 and 90

d. 20 and 60

e. 50 and 12

f. 70 and 55

g. 24 and 18

h. 60 and 16

i. 72 and 30

j. 20 and 12

Part 2h: HCF of two numbers (Solutions)

Find the HCF of the following numbers

a. 100 and 45

The factors of 100 are 1,2,4,5,10,20,25,50,100
The factors of 45 are 1,3,5,9,15,45
So, the HCF is 5

b. 40 and 54

The factors of 40 are 1,2,4,5,8,10,20,40
The factors of 54 are 1,2,3,6,9,18,27,54
So, the HCF is 2

c. 74 and 90

The factors of 74 are 1,2,37,74
The factors of 90 are 1,2,3,5,6,9,10,15,30,45,90
So, the HCF is 2

d. 20 and 60

The factors of 20 are 1,2,4,5,10,20
The factors of 60 are 1,2,3,4,5,6,10,12,15,20,30,60
So, the HCF is 20

e. 50 and 12

The factors of 50 are 1,2,5,10,25,50
The factors of 12 are 1,2,3,4,6,12
So, the HCF is 2

f. 70 and 55

The factors of 70 are 1,2,5,7,10,14,35,70
The factors of 55 are 1,5,11,55
So, the HCF is 5

g. 24 and 18

The factors of 24 are 1,2,3,4,6,8,12,24
The factors of 18 are 1,2,3,9,18
So, the HCF is 3

h. 60 and 16

The factors of 60 are 1,2,3,4,5,6,10,12,15,20,30,60
The factors of 16 are 1,2,4,8,16
So, the HCF is 4

i. 72 and 30

The factors of 72 are 1,2,3,4,6,8,9,12,24,36,72
The factors of 30 are 1,2,3,5,6,10,15,30
So, the HCF is 6

j. 20 and 12

The factors of 20 are 1,2,4,5,10,20
The factors of 12 are 1,2,3,4,6,12

So, the HCF is 4

Part 2i: HCF of two numbers

(Questions)

Find the HCF of the following numbers

a. 70 and 52

b. 20 and 88

c. 60 and 20

d. 96 and 28

e. 45 and 25

f. 74 and 28

g. 30 and 90

h. 50 and 48

i. 80 and 16

j. 72 and 15

Part 2i: HCF of two numbers

(Solutions)

Find the HCF of the following numbers

a. 70 and 52

The factors of 70 are 1,2,5,7,10,14,35,70
The factors of 52 are 1,2,4,13,26,52
So, the HCF is 2

b. 20 and 88

The factors of 20 are 1,2,4,5,10,20
The factors of 88 are 1,2,4,8,11,22, ...
So, the HCF is 4

c. 60 and 20

The factors of 60 are 1,2,3,4,5,6,10,12,15,20,30,60
The factors of 20 are 1,2,4,5,10,20
So, the HCF is 20

d. 96 and 28

The factors of 96 are 1,2,3,4,6,12,16,24,32,48,96
The factors of 28 are 1,2,4,7,14,28
So, the HCF is 4

e. 45 and 25

The factors of 45 are 1,3,5,9,15,45
The factors of 25 are 1,5,25
So, the HCF is 5

f. 74 and 28

The factors of 74 are 1,2,37,74
The factors of 28 are 1,2,4,7,14,28
So, the HCF is 2

g. 30 and 90

The factors of 30 are 1,2,3,5,6,10,15,30
The factors of 90 are 1,2,3,5,6,9,10,15,30,45,90
So, the HCF is 30

h. 50 and 48

The factors of 50 are 1,2,5,10,25,50
The factors of 48 are 1,2,3,4,6,8,12,16,24,48
So, the HCF is 2

i. 80 and 16

The factors of 80 are 1,2,4,5,8,10,16,20,40,80
The factors of 16 are 1,2,4,8,16
So, the HCF is 16

j. 72 and 15

The factors of 72 are 1,2,3,4,6,8,9,12,24,36,72
The factors of 15 are 1,3,5,15

So, the HCF is 3

Part 2j: HCF of two numbers

(Questions)

Find the HCF of the following numbers

a. 72 and 6

b. 60 and 4

c. 30 and 4

d. 100 and 25

e. 50 and 8

f. 80 and 20

g. 40 and 18

h. 36 and 12

i. 4 and 44

j. 7 and 28

Part 2j: HCF of two numbers

(Solutions)

Find the HCF of the following numbers

a. 72 and 6

The factors of 72 are 1,2,3,4,6,8,9,12,24,36,72
The factors of 6 are 1,2,3,6
So, the HCF is 6

b. 60 and 4

The factors of 60 are 1,2,3,4,5,6,10,12,15,20,30,60
The factors of 4 are 1,2,4
So, the HCF is 4

c. 30 and 4

The factors of 30 are 1,2,3,5,6,10,15,30
The factors of 4 are 1,2,4
So, the HCF is 2

d. 100 and 25

The factors of 100 are 1,2,4,5,10,20,25,50,100
The factors of 25 are 1,5,25
So, the HCF is 25

e. 50 and 8

The factors of 50 are 1,2,5,10,25,50
The factors of 8 are 1,2,4,8
So, the HCF is 2

f. 80 and 20

The factors of 80 are 1,2,4,5,8,10,16,20,40,80
The factors of 20 are 1,2,4,5,10,20
So, the HCF is 20

g. 40 and 18

The factors of 40 are 1,2,4,5,8,10,20,40
The factors of 18 are 1,2,3,6,9,18
So, the HCF is 2

h. 36 and 12

The factors of 36 are 1,2,3,4,6,9,12,18,36
The factors of 12 are 1,2,3,4,6,12
So, the HCF is 12

i. 4 and 44

The factors of 4 are 1,2,4
The factors of 44 are 1,2,4,11,.....
So, the HCF is 4

j. 7 and 28

The factors of 7 are 1,7
The factors of 28 are 1,2,4,7,14,28

So, the HCF is 7

Part 3a: HCF of three numbers (Questions)

Find the HCF of the following numbers

a. 30, 50, 75

b. 20, 25, 30

c. 22, 28, 32

d. 40, 45, 46

e. 18, 22, 24

f. 22, 26, 34

g. 21, 30, 32

h. 32, 46, 52

i. 34, 36, 60

j. 10, 26, 28

Part 3a: HCF of three numbers (Solutions)

Find the HCF of the following numbers

a. 30, 50, 75

The factors of 30 are 1,2,3,5,6,10,15,30
The factors of 50 are 1,2,5,10,25,50
The factors of 75 are 1,3,5,15,25,75
So, the HCF is 5

b. 20, 25, 30

The factors of 20 are 1,2,4,5,10,20
The factors of 25 are 1,5,25
The factors of 30 are 1,2,3,5,6,10,15,30
So, the HCF is 5

c. 22, 28, 32

The factors of 22 are 1,2,11,22
The factors of 28 are 1,2,4,7,14,28
The factors of 32 are 1,2,4,8,16,32
So, the HCF is 2

d. 40, 45, 46

The factors of 40 are 1,2,4,5,8,10,20,40
The factors of 45 are 1,3,5,9,15,45
The factors of 46 are 1,2,23,46
So, the HCF is 1

e. 18, 22, 24

The factors of 18 are 1,2,3,6,9,18
The factors of 22 are 1,2,11,22
The factors of 24 are 1,2,3,4,6,8,12,24
So, the HCF is 2

f. 22, 26, 34

The factors of 22 are 1,2,11,22
The factors of 26 are 1,2,13,26
The factors of 34 are 1,2,17,34
So, the HCF is 2

g. 21, 30, 32

The factors of 21 are 1,3,7,21
The factors of 30 are 1,2,3,5,6,10,15,30
The factors of 32 are 1,2,4,8,16,32
So, the HCF is 1

h. 32, 46, 52

The factors of 32 are 1,2,4,8,16,32
The factors of 46 are 1,2,23,46
The factors of 52 are 1,2,4,13,26,52
So, the HCF is 2

i. 34, 36, 60

The factors of 34 are 1,2,17,34
The factors of 36 are 1,2,3,4,6,9,12,18,36
The factors of 60 are 1,2,3,4,5,6,10,12,15,20,30,60
So, the HCF is 2

j. 10,26,28

The factors of 10 are 1,2,5,10
The factors of 26 are 1,2,13,26
The factors of 28 are 1,2,4,7,14,28
So, the HCF is 2

Part 3b: HCF of three numbers (Questions)

Find the HCF of the following numbers

a. 20, 27, 34
b. 50, 54, 60
c. 60, 70, 90
d. 10, 60, 72
e. 50, 52, 56
f. 60, 62, 90
g. 70, 78, 80
h. 50, 52, 100
i. 70, 80, 84
j. 70, 84, 90

Part 3b: HCF of three numbers (Solutions)

Find the HCF of the following numbers

a. 20, 27, 34

The factors of 20 are 1,2,4,5,10,20
The factors of 27 are 1,3,9,27
The factors of 34 are 1,2,17,34
So, the HCF is 1

b. 50, 54, 60

The factors of 50 are 1,2,5,10,25,50
The factors of 54 are 1,2,3,18,27,54
The factors of 60 are 1,2,3,4,5,6,10,12,15,20,30,60
So, the HCF is 2

c. 60, 70, 90

The factors of 60 are 1,2,3,4,5,6,10,12,15,20,30,60
The factors of 70 are 1,2,5,7,10,14,35,70
The factors of 90 are 1,2,3,5,6,9,10,15,30,45,90
So, the HCF is 10

d. 10, 60, 72

The factors of 10 are 1,2,5,10
The factors of 60 are 1,2,3,4,5,6,10,12,15,20,30,60
The factors of 72 are 1,2,3,4,6,8,9,12,24,36,72
So, the HCF is 2

e. 50, 52, 56

The factors of 50 are 1,2,5,10,25,50
The factors of 52 are 1,2,4,13,26,52
The factors of 56 are 1,2,4,14,28,56
So, the HCF is 2

f. 60, 62, 90

The factors of 60 are 1,2,3,4,5,6,10,12,15,20,30,60
The factors of 62 are 1,2,31,62
The factors of 90 are 1,2,3,5,6,9,10,15,30,45,90
So, the HCF is 2

g. 70, 78, 80

The factors of 70 are 1,2,5,7,10,14,35,70
The factors of 78 are 1,2,3,6,13,26,39,78
The factors of 80 are 1,2,4,5,8,10,16,20,40,80
So, the HCF is 2

h. 50, 52, 100

The factors of 50 are 1,2,5,10,25,50
The factors of 52 are 1,2,4,13,26,52
The factors of 100 are 1,2,4,5,10,20,25,50,100
So, the HCF is 2

i. 70, 80, 84

The factors of 70 are 1,2,5,7,10,14,35,70
The factors of 80 are 1,2,4,5,8,10,16,20,40,80
The factors of 84 are 1,2,3,4,6,7,12,14,21,28,42,84
So, the HCF is 2

j. 70, 84, 90

The factors of 70 are 1,2,5,7,10,14,35,70
The factors of 84 are 1,2,3,4,6,7,12,14,21,28,42,84
The factors of 90 are 1,2,3,5,6,9,10,15,30,45,90
So, the HCF is 2

Part 3c: HCF of three

numbers (Questions)

Find the HCF of the following numbers

a. 12, 45, 70

b. 16, 33, 90

c. 8, 10, 70

d. 14, 16, 40

e. 10, 12, 16

f. 4, 8, 14

g. 6, 8, 18

h. 16, 20, 24

i. 4, 9, 18

j. 8,12,16

Part 3c: HCF of three numbers (Solutions)

Find the HCF of the following numbers

a. 12, 45, 70

The factors of 12 are 1,2,3,4,6,12
The factors of 45 are 1,3,5,9,15,45
The factors of 70 are 1,2,5,7,10,14,70
So, the HCF is 1

b. 16, 33, 90

The factors of 16 are 1,2,4,8,16
The factors of 33 are 1,3,11,33
The factors of 90 are 1,2,3,5,6,9,10,15,30,45,90
So, the HCF is 1

c. 8, 10, 70

The factors of 8 are 1,2,4,8
The factors of 10 are 1,2,5,10
The factors of 70 are 1,2,5,7,10,14,35,70
So, the HCF is 2

d. 14, 16, 40

The factors of 14 are 1,2,7,14
The factors of 16 are 1,2,4,8,16
The factors of 40 are 1,2,4,5,8,10,20,40
So, the HCF is 2

e. 10, 12, 16

The factors of 10 are 1,2,5,10
The factors of 12 are 1,2,3,4,6,12
The factors of 16 are 1,2,4,8,16
So, the HCF is 2

f. 4, 8, 14

The factors of 4 are 1,2,4
The factors of 8 are 1,2,4,8
The factors of 14 are 1,2,7,14
So, the HCF is 2

g. 6, 8, 18

The factors of 6 are 1,2,3,6
The factors of 8 are 1,2,4,8
The factors of 18 are 1,2,3,6,9,18
So, the HCF is 2

h. 16, 20, 24

The factors of 16 are 1,2,4,8,16
The factors of 20 are 1,2,4,5,10,20
The factors of 24 are 1,2,3,4,6,8,12,24
So, the HCF is 4

i. 4, 9, 18

The factors of 4 are 1,2,4
The factors of 9 are 1,3,9
The factors of 24 are 1,2,3,4,6,8,12,24
So, the HCF is 1

j. 8,12,16

The factors of 8 are 1,2,4,8
The factors of 12 are 1,2,3,4,6,12
The factors of 16 are 1,2,4,8,16
So, the HCF is 4

Part 3d: HCF of three numbers (Questions)

Find the HCF of the following numbers

a. 6, 9, 10
b. 10, 14, 32
c. 50, 55, 66
d. 12, 18, 72
e. 7, 24, 35
f. 40, 45, 100
g. 25, 50, 15
h. 18, 20, 45
i. 25, 40, 75
j. 9, 20,100

Part 3d: HCF of three numbers (Solutions)

Find the HCF of the following numbers

a. 6, 9, 10

The factors of 6 are 1,2,3,6
The factors of 9 are 1,3,9
The factors of 10 are 1,2,5,10
So, the HCF is 1

b. 10, 14, 32

The factors of 10 are 1,2,5,10
The factors of 14 are 1,2,7,14
The factors of 32 are 1,2,4,8,16,32
So, the HCF is 2

c. 50, 55, 66

The factors of 50 are 1,2,5,10,25,50
The factors of 55 are 1,5,11,55
The factors of 66 are 1,2,3,6,11,22,33,66
So, the HCF is 1

d. 12, 18, 72

The factors of 12 are 1,2,3,4,6,12
The factors of 18 are 1,2,3,6,9,18
The factors of 72 are 1,2,3,4,6,8,9,12,18,24,36,72
So, the HCF is 6

e. 7, 24, 35

The factors of 7 are 1,7
The factors of 24 are 1,2,3,4,6,8,12,24
The factors of 35 are 1,5,7,35
So, the HCF is 1

f. 40, 45, 100

The factors of 40 are 1,2,4,5,8,10,20,40
The factors of 45 are 1,3,5,9,15,45
The factors of 100 are 1,2,4,5,10,20,25,50,100
So, the HCF is 5

g. 25, 50, 15

The factors of 25 are 1,5,25
The factors of 50 are 1,2,5,10,25,50
The factors of 15 are 1,3,5
So, the HCF is 5

h. 18, 20, 45

The factors of 18 are 1,2,3,6,9,18
The factors of 20 are 1,2,4,5,10,20
The factors of 45 are 1,3,5,9,15,45
So, the HCF is 1

i. 25, 40, 75

The factors of 25 are 1,5,25
The factors of 40 are 1,2,4,5,8,10,20,40
The factors of 75 are 1,3,5,15,25,75
So, the HCF is 5

j. 9,20,100

The factors of 9 are 1,3,9
The factors of 20 are 1,2,4,5,10,20
The factors of 100 are 1,2,4,5,10,20,25,50,100
So, the HCF is 1

Part 3e: HCF of three numbers (Questions)

Find the HCF of the following numbers

a. 7, 14, 15
b. 10, 18, 27
c. 50, 55, 66
d. 6, 7, 35
e. 9, 12, 40
f. 20, 24, 30
g. 2, 12, 18
h. 8, 9, 27
i. 7, 77, 88
j. 15, 50, 35

Part 3e: HCF of three numbers (Solutions)

Find the HCF of the following numbers

a. 7, 14, 15

The factors of 7 are 1,7
The factors of 14 are 1,2,7,14
The factors of 15 are 1,3,5,15
So, the HCF is 1

b. 10, 18, 27

The factors of 10 are 1,2,5,10
The factors of 18 are 1,2,3,6,9,18
The factors of 27 are 1,3,9,27
So, the HCF is 1

c. 50, 55, 66

The factors of 50 are 1,2,5,10,25,50
The factors of 1,5,11,55
The factors of 66 are 1,2,3,6,11,22,33,66
So, the HCF is 1

d. 6, 7, 35

The factors of 6 are 1,2,3,6
The factors of 7 are 1,7
The factors of 35 are 1,5,7,35
So, the HCF is 1

e. 9, 12, 40

The factors of 9 are 1,3,9
The factors of 12 are 1,2,3,4,6,12
The factors of 40 are 1,2,4,5,8,10,20,40
So, the HCF is 1

f. 20, 24, 30

The factors of 20 are 1,2,4,5,10,20
The factors of 24 are 1,2,3,4,6,8,12,24
The factors of 30 are 1,2,3,5,6,10,15,30
So, the HCF is 2

g. 2, 12, 18

The factors of 2 are 1,2
The factors of 12 are 1,2,3,4,6,12
The factors of 18 are 1,2,3,6,9,18
So, the HCF is 2

h. 8, 9, 27

The factors of 8 are 1,2,4,8
The factors of 9 are 1,3,9
The factors of 27 are 1,3,9,27
So, the HCF is 1

i. 7, 77, 88

The factors of 7 are 1,7
The factors of 77 are 1,7,11,77
The factors of 88 are 1,8,11,44,88
So, the HCF is 1

j. 15, 50, 35

The factors of 15 are 1,3,5,15
The factors of 50 are 1,2,5,10,25,50
The factors of 35 are 1,5,7,35
So, the HCF is 5

Part 3f: HCF of three numbers (Questions)

Find the HCF of the following numbers

a. 14, 50, 70
b. 18, 20, 30
c. 8, 12, 40
d. 22, 32, 72
e. 50, 100, 75
f. 10, 40, 50
g. 35, 45, 50
h. 32, 40, 42
i. 60, 66, 100
j. 9, 10, 100

Part 3f: HCF of three

numbers (Solutions)

Find the HCF of the following numbers

a. 14, 50, 70

The factors of 14 are 1,2,7,14
The factors of 50 are 1,2,5,10,25,50
The factors of 70 are 1,2,5,7,10,14,35,70
So, the HCF is 2

b. 18, 20, 30

The factors of 18 are 1,2,3,6,9,18
The factors of 20 are 1,2,5,10,20
The factors of 30 are 1,2,5,6,10,15,30
So, the HCF is 2

c. 8, 12, 40

The factors of 8 are 1,2,4,8
The factors of 12 are 1,2,3,4,6,12
The factors of 40 are 1,2,4,5,8,10,20,40
So, the HCF is 4

d. 22, 32, 72

The factors of 22 are 1,2,11,22
The factors of 32 are 1,2,4,8,16,32
The factors of 72 are 1,2,3,4,6,8,9,12,18,24,36,72
So, the HCF is 2

e. 50, 100, 75

The factors of 50 are 1,2,5,10,25,50
The factors of 100 are 1,2,4,5,10,20,25,50,100
The factors of 75 are 1,3,5,15,25,75
So, the HCF is 25

f. 10, 40, 50

The factors of 10 are 1,2,5,10
The factors of 40 are 1,2,4,5,8,10,20,40
The factors of 50 are 1,2,5,10,25,50
So, the HCF is 10

g. 35, 45, 50

The factors of 35 are 1,5,7,35
The factors of 45 are 1,3,5,9,15,45
The factors of 50 are 1,2,5,10,25,50
So, the HCF is 5

h. 32, 40, 42

The factors of 32 are 1,2,4,8,16,32
The factors of 40 are 1,2,4,5,8,10,20,40
The factors of 42 are 1,2,3,6,7,14,21,42
So, the HCF is 2

i. 60, 66, 100

The factors of 60 are 1,2,3,4,5,6,10,12,15,20,30,60
The factors of 66 are 1,2,3,6,11,22,33,66
The factors of 100 are 1,2,4,5,10,20,25,50,100
So, the HCF is 2

j. 9,10,100

The factors of 9 are 1,3,9
The factors of 10 are 1,2,5,10
The factors of 100 are 1,2,4,5,10,20,25,50,100
So, the HCF is 1

Part 3G: HCF of three numbers (Questions)

Find the HCF of the following numbers

a. 2, 8, 10

b. 9, 18, 24

c. 8, 14, 21

d. 20, 30, 55

e. 24, 28, 80

f. 15, 22, 50

g. 10, 25, 75

h. 10, 50, 100

i. 25, 55, 60

j. 12,14, 24

Part 3G: HCF of three numbers (Solutions)

Find the HCF of the following numbers

a. 2, 8, 10

The factors of 2 are 1,2
The factors of 8 are 1,2,4,8
The factors of 10 are 1,2,5,10
So, the HCF is 2

b. 9, 18, 24

The factors of 9 are 1,3,9
The factors of 18 are 1,2,3,6,9,18
The factors of 24 are 1,2,3,4,6,8,12,24
So, the HCF is 3

c. 8, 14, 21

The factors of 8 are 1,2,4,8
The factors of 14 are 1,2,7,14
The factors of 21 are 1,3,7,21
So, the HCF is 1

d. 20, 30, 55

The factors of 20 are 1,2,4,5,10,20
The factors of 30 are 1,2,3,5,6,10,15,30
The factors of 55 are 1,5,11,55
So, the HCF is 5

e. 24, 28, 80

The factors of 24 are 1,2,3,4,6,8,12,24
The factors of 28 are 1,2,4,7,14,28
The factors of 80 are 1,2,4,5,8,10,16,20,40,80
So, the HCF is 4

f. 15, 22, 50

The factors of 15 are 1,3,5,15
The factors of 22 are 1,2,11,22
The factors of 50 are 1,2,5,10,25,50
So, the HCF is 1

g. 10, 25, 75

The factors of 10 are 1,2,5,10
The factors of 25 are 1,5,25
The factors of 75 are 1,3,5,15,25,75
So, the HCF is 5

h. 10, 50, 100

The factors of 10 are 1,2,5,10
The factors of 50 are 1,2,5,10,25,50
The factors of 100 are 1,2,4,5,10,20,25,50,100
So, the HCF is 10

i. 25, 55, 60

The factors of 25 are 1,5,25
The factors of 55 are 1,5,11,55
The factors of 60 are 1,2,3,4,5,6,10,12,15,20,30,60
So, the HCF is 5

j. 12,14,24

The factors of 12 are 1,2,3,4,6,12
The factors of 14 are 1,2,7,14
The factors of 24 are 1,2,3,4,6,8,12,24
So, the HCF is 2

Part 3h: HCF of three numbers (Questions)

Find the HCF of the following numbers

a. 15, 72, 100
b. 20, 60, 75
c. 12, 75, 100
d. 22, 24, 100
e. 8, 16, 25
f. 6, 12, 32
g. 12, 14, 16
h. 15, 25, 35
i. 8, 18, 100
j. 10,15, 35

Part 3h: HCF of three

numbers (Solutions)

Find the HCF of the following numbers

a. 15, 72, 100

The factors of 15 are 1,3,5,15
The factors of 72 are 1,2,3,4......72
The factors of 100 are 1,2,4,5,10,20,25,50,100
So, the HCF is 1

b. 20, 60, 75

The factors of 20 are 1,2,4,5,10,20
The factors of 60 are 1,2,3,4,5,6,10,12,15,20,30,60
The factors of 75 are 1,3,5,15,25,75
So, the HCF is 5

c. 12, 75, 100

The factors of 12 are 1,2,3,4,6,12
The factors of 75 are 1,3,5,15,25,75
The factors of 100 are 1,2,4,5,10,20,25,50,100
So, the HCF is 1

d. 22, 24, 100

The factors of 22 are 1,2,11,22
The factors of 24 are 1,2,3,4,6,8,12,24
The factors of 100 are 1,2,4,5,10,20,25,50,100
So, the HCF is 2

e. 8, 16, 25

The factors of 8 are 1,2,4,8
The factors of 16 are 1,2,4,8,16
The factors of 25 are 1,5,25
So, the HCF is 1

f. 6, 12, 32

The factors of 6 are 1,2,3,6
The factors of 12 are 1,2,3,4,6,12
The factors of 32 are 1,2,4,8,16,32
So, the HCF is 2

g. 12, 14, 16

The factors of 12 are 1,2,3,4,6,12
The factors of 14 are 1,2,7,14
The factors of 16 are 1,2,4,8,16
So, the HCF is 2

h. 15, 25, 35

The factors of 15 are 1,3,5,15
The factors of 25 are 1,5,25
The factors of 35 are 1,5,7,35
So, the HCF is 5

i. 8, 18, 100

The factors of 8 are 1,2,4,8
The factors of 18 are 1,2,6,9,18
The factors of 100 are 1,2,4,5,10,20,25,50,100
So, the HCF is 2

j. 10,15,35

The factors of 10 are 1,2,5,10
The factors of 15 are 1,3,5,15
The factors of 35 are 1,5,7,35
So, the HCF is 5

Part 3i: HCF of three numbers (Questions)

Find the HCF of the following numbers

a. 12, 18, 50
b. 14, 16, 60
c. 12, 15, 80
d. 12, 15, 21
e. 14, 50, 100
f. 22, 24, 32
g. 8, 12, 20
h. 15, 50,100
i. 10, 22, 36
j. 8, 22, 24

Part 3i: HCF of three numbers (Solutions)

Find the HCF of the following numbers

a. 12, 18, 50

The factors of 12 are 1,2,3,4,6,12
The factors of 18 are 1,2,3,6,9,18
The factors of 50 are 1,2,5,10,25,50
So, the HCF is 2

b. 14, 16, 60

The factors of 14 are 1,2,7,14
The factors of 16 are 1,2,4,8,16
The factors of 60 are 1,2,3,4,5,6,10,12,15,20,30,60
So, the HCF is 2

c. 12, 15, 80

The factors of 12 are 1,2,3,4,6,12
The factors of 15 are 1,3,5,15
The factors of 80 are 1,2,4,5,8,10,16,20,40,80
So, the HCF is 1

d. 12, 15, 21

The factors of 12 are 1,2,3,4,6,12
The factors of 15 are 1,3,5,15
The factors of 21 are 1,3,7,21
So, the HCF is 3

e. 14, 50, 100

The factors of 14 are 1,2,7,14
The factors of 50 are 1,2,5,10,25,50
The factors of 100 are 1,2,4,5,10,20,25,50,100
So, the HCF is 2

f. 22, 24, 32

The factors of 22 are 1,2,11,22
The factors of 24 are 1,2,3,4,6,8,12,24
The factors of 32 are 1,2,4,8,16,32
So, the HCF is 2

g. 8, 12, 20

The factors of 8 are 1,2,4,8
The factors of 12 are 1,2,3,4,6,12
The factors of 20 are 1,2,4,5,10,20
So, the HCF is 4

h. 15, 50,100

The factors of 15 are 1,3,5,15
The factors of 50 are 1,2,5,10,25,50
The factors of 100 are 1,2,4,5,10,20,25,50,100
So, the HCF is 5

i. 10, 22,36

The factors of 10 are 1,2,5,10
The factors of 22 are 1,2,11,22
The factors of 36 are 1,2,3,4,6,9,12,18,36
So, the HCF is 2

j. 8,22,24

The factors of 8 are 1,2,4,8
The factors of 22 are 1,2,11,22
The factors of 24 are 1,2,3,4,6,8,12,24
So, the HCF is 2

Part 3j: HCF of three numbers (Questions)

Find the HCF of the following numbers

a. 4, 6, 100

b. 4, 8, 24

c. 8, 20, 40

d. 22, 80, 90

e. 24, 66, 100

f. 22, 32, 42

g. 14, 32, 60

h. 14, 24, 50

i. 5, 25, 50

j. 10, 72, 80

Part 3j: HCF of three numbers (Solutions)

Find the HCF of the following numbers

a. 4, 6, 100

The factors of 4 are 1,2,4
The factors of 6 are 1,2,3,6
The factors of 100 are 1,2,4,5,10,20,25,50,100
So, the HCF is 2

b. 4, 8, 24

The factors of 4 are 1,2,4
The factors of 8 are 1,2,4,8
The factors of 24 are 1,2,3,4,6,8,12,24
So, the HCF is 4

c. 8, 20, 40

The factors of 8 are 1,2,4,8
The factors of 20 are 1,2,4,5,10,20
The factors of 40 are 1,2,4,5,8,10,20,40
So, the HCF is 4

d. 22, 80, 90

The factors of 22 are 1,2,11,22
The factors of 80 are 1,2,4,5,8,10,16,20,40,80
The factors of 90 are 1,2,3,5,6,9,10,15,30,45,90
So, the HCF is 2

e. 24, 66, 100

The factors of 24 are 1,2,3,4,6,8,12,24
The factors of 66 are 1,2,3,6,11,22,33,66
The factors of 100 are 1,2,4,5,10,20,25,50,100
So, the HCF is 2

f. 22, 32, 42

The factors of 22 are 1,2,11,22
The factors of 32 are 1,2,4,8,16,32
The factors of 42 are 1,2,3,6,7,14,21,42
So, the HCF is 2

g. 14, 32, 60

The factors of 14 are 1,2,7,14
The factors of 32 are 1,2,4,8,16,32
The factors of 60 are 1,2,3,4,5,6,10,12,15,20,30,60
So, the HCF is 2

h. 14, 24,50

The factors of 14 are 1,2,7,14
The factors of 24 are 1,2,3,4,6,8,12,24
The factors of 50 are 1,2,5,10,25,50
So, the HCF is 2

i. 5, 25,50

The factors of 5 are 1,5
The factors of 25 are 1,5,25
The factors of 50 are 1,2,5,10,25,50
So, the HCF is 5

j. 10,72,80

The factors of 10 are 1,5,10
The factors of 72 are 1,2,3,4,6,8,9,12,18,24,36,72
The factors of 80 are 1,2,4,5,8,10,16,20,40,80
So, the HCF is 2

Conclusion

Thank you once again for purchasing this book. I hope it has helped you in your journey to understand HCF.

Please, if you learnt something from this book, I would like you to leave a review. It'd be appreciated.

Thank you.

www.ingramcontent.com/pod-product-compliance
Lightning Source LLC
Chambersburg PA
CBHW070608220526
45467CB00003B/1341